# Our Changing World

*Navigating Change in a Sustainable, Interconnected World*

Series Synopsis

James Fountain

James Fountain

TREELINE PUBLICATIONS

**ISBNs:**

Paperback: 978-1-963443-16-5
eBook: 978-1-963443-17-2

# Table of Contents

# Acknowledgments And Dedication

Before we venture into the heart of this series, let's take a moment to acknowledge the hands and hearts that have been instrumental in bringing this work to life. It is with deep gratitude that I recognize the individuals and communities whose unwavering support and boundless inspiration have laid the cornerstone for these pages. Their contributions have not only enriched this journey but have also been pivotal in shaping the narratives that unfold within.

To start, this series stands as a testament to the guardians of wisdom through the ages—Indigenous peoples, whose very essence is woven into the Earth's rhythms, educators and seekers of knowledge who illuminate our understanding, and you, the valiant souls navigating the sustainability frontier. Your endeavors, though often stretched thin by the magnitude of your obligations and the ceaseless demands for your expertise, are the cornerstone upon which we dare to envision a future brimming with hope and harmony. In these pages lies a shared reservoir of insights, a lighthouse for those teetering on the edge of this crucial movement, aiming to cast light on the path ahead with lessons from my journey, igniting a spark within yours.

The dedication you exhibit, even when overwhelmed and overcommitted, underscores a collective resolve that transcends personal accolades. It's driven by a deep-seated desire for impact, a shared obsession with catalyzing meaningful change in the realms of sustainability, climate action, and human rights. Your commitment speaks to a profound yearning for a deeper connection with the world around us, a connection that is less about individual recognition and more about the collective soul of our planet. In an era shadowed by division, your tireless efforts weave a narrative of unity and hope, showcasing the monumental impact we can achieve through values-driven action and the pursuit of real, tangible solutions. This series seeks not only to honor your undertaking but to amplify it, offering it as a clarion call to all who dream of a tomorrow where our planet is revered, and our collective well-being is cherished.

In this vast journey, Mona, your steadfast support has been my compass, guiding me through the tumultuous and the tranquil alike. Your sacrifices, quietly monumental, from the solitary evenings while I was ensconced in thought or sequestered at the local bar, to the boundless reservoir of patience and encouragement you offered amid my sea of doubts. Our countless brainstorming sessions—naming series, books, and chapters—have been pivotal. This endeavor, imbued with your love and belief, mirrors not just my efforts but the essence of your spirit. The laughter we shared, our nocturnal dialogues, and even the moments of companionable silence have woven a rich tapestry into this work. You have been my muse, my confidante, and my untiring partner through every zenith and nadir. Your love, steadfast and grounding; your incredible life story, a wellspring of inspiration; this milestone bears witness to the enduring strength of our shared journey.

As we stand on the cusp of new adventures, I am buoyed by the knowledge that with you by my side, there are no bounds to the stories we can tell, the worlds we can explore, and the impacts we can make. Here's to the chapters yet unwritten, the tales yet untold, and the journey that continues to unfold with you, Mona, as my guiding star.

And finally, to Liam, my cherished companion whose absence has left a silence too profound for words. I miss you, buddy. In my heart, I envision you in a realm where the forest trails stretch into eternity, where mountain meadows sprawl under the open sky, where tennis balls are abundant, and marrow-filled bovine bones are yours for the taking.

Liam, you exemplified the essence of an ideal Earth citizen—your love, dedication, empathy, and compassion were lessons in themselves, showing me the heights to which we can aspire. Your soul, luminous and guiding, illuminated the path of what life should embody. Through your eyes, I learned to see the world not just as it is but as it could be, filled with love and boundless joy. This book and the series that contains it, while a dedication to the guardians of our planet and the stewards of knowledge, is also a tribute to you, Liam. Your spirit, a guiding star in the vast cosmos of our journey, continues to lead me toward love, empathy, and the profound connections that weave the fabric of life. Thank you for teaching me, for being my compass in exploring what it truly means to live fully and love unconditionally. Your legacy is etched not just in these pages but in my ambition of the being I hope to become.

-JF-

James Fountain

# Our Changing World

*Navigating Change in a Sustainable, Interconnected World*

## Threads of Green

Weaving Sustainability
into the Cultural Fabric

## A Sustainable Civilization

Unraveling the
Threads of Change

## A Planet in Balance

Exploring the Socio-Cultural
Landscape of Sustainability

## The Responsibility Renaissance

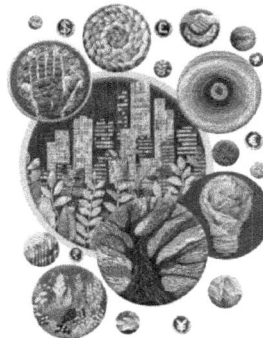

Business as a Catalyst for
Environmental and Social Ethics

# Our Changing World: Navigating Change in a Sustainable, Interconnected World

Our Changing World is a comprehensive exploration of sustainability, seamlessly blending cultural, ethical, socio-economic, and business perspectives into a unified narrative. This series delves deeply into how diverse cultural backgrounds and traditional knowledge systems are integral to understanding and addressing global sustainability challenges. By exploring the rich tapestry of global cultural practices alongside modern scientific advancements, the series underscores the importance of integrating systems thinking with a broad array of traditional and contemporary insights to tackle issues like environmental degradation and social resilience in the Anthropocene.

The narrative evolves to emphasize the historical roots of sustainable thought and its crucial applications in today's world, addressing pressing issues such as climate change, renewable energy, and biodiversity. It highlights the interconnectedness of human actions and environmental impacts, showcasing how understanding our past can inform present actions and future strategies. The discourse then shifts to explore the ethical and socio-cultural dimensions of sustainability, stressing the need for a robust ethical framework that includes environmental justice, social equity, and human rights. It illustrates how education, art, and storytelling are powerful tools for raising awareness and motivating societal change, bridging the gap between knowledge and action.

As the discussion transitions into the realm of business, the series illustrates how sustainability has become a critical business imperative. It explores how businesses can drive substantial change by adopting regenerative, waste-minimizing, and resource-efficient practices, and how integrating Environmental, Social, and Governance (ESG) factors into business operations is essential for sustainable development. Leadership's role in fostering a culture of sustainability within the corporate sector is also highlighted, emphasizing how ethical and innovative business practices can influence global sustainability efforts.

Our Changing World serves as a call to action for individuals, communities, businesses, and policymakers to leverage the power of cultural diversity, ethical imperatives, and innovative business practices to forge a sustainable future. This series is an indispensable resource for anyone seeking to understand the complex interplay between culture, ethics, business, and sustainability, providing a holistic view that encourages all sectors of society to contribute to environmental stewardship and sustainable development.

James Fountain

# Threads of Green

## Weaving Sustainability into the Cultural Fabric

## CHAPTER SYNOPSIS

### James Fountain

Our Changing World

# Book 1: Threads of Green
*Weaving Sustainability
into the Cultural Fabric*

# Threads of Green: Weaving Sustainability into the Cultural Fabric

*Threads of Green: Weaving Sustainability into the Cultural Fabric*, the first book in the *Our Changing World* series, establishes a critical understanding of how cultural diversity and unity are essential to sustainability. It explores how different cultures across the globe perceive and approach sustainability challenges, emphasizing the integration of systems thinking with both traditional and contemporary knowledge to tackle issues like environmental degradation and social resilience in the Anthropocene. This approach highlights the importance of understanding diverse cultural perspectives to appreciate the interconnectedness that is crucial for addressing sustainability on various fronts.

Positioned at the intersection of cultural studies and environmental science, the book argues that a sustainable future requires a deep appreciation of traditional knowledge alongside modern scientific advancements. By tracing how cultural landscapes have evolved, Threads of Green demonstrates how traditional practices and beliefs provide unique and effective solutions to today's ecological challenges.

The book emphasizes the role of cultural diversity in enhancing the global sustainability conversation and shows how global and local collaborative efforts are vital for effective environmental stewardship. Through narratives of resilience, innovation, and collective action, it illustrates how a synthesis of historical insights and modern practices is key to developing sustainable solutions.

The narrative concludes with an examination of cultural case studies which explores historical and time-tested sustainability practices from remote communities worldwide. From the ancient terraced fields in the

Philippines to the ice-covered landscapes of the Arctic, these case studies demonstrate how local knowledge and traditional practices play a crucial role in global sustainability efforts. Highlighting practices such as the Ifugao Rice Terraces in the Philippines, the sustainable hunting techniques of the Inuit, and the ancient water management systems of the Qanat in Iran, this section offers valuable insights into how cultural heritage can harmonize with environmental stewardship.

Overall, Threads of Green serves as a call to action for individuals, communities, and policymakers to leverage the power of cultural diversity and integrated knowledge to forge a sustainable future. This book is an essential guide for anyone seeking to understand the complex relationship between culture and sustainability and provides a comprehensive perspective on integrating these insights into the fabric of everyday life.

# Preface

The preface introduces the book's theme of navigating the complexities of ecological sustainability with hope and determination. It emphasizes the urgency of addressing ecological challenges amid rapid changes, drawing on a broad array of perspectives—from scientists to activists—to guide readers toward a sustainable future. The author stresses the importance of learning from both historical and current contexts to shape effective sustainability strategies and invites readers to actively engage and contribute to these discussions.

# Introduction:
# Weaving a Global Tapestry

The introduction of the book, *Weaving a Global Tapestry*, explores the idea of weaving a diverse global tapestry that integrates the unique cultural narratives from around the world with a commitment to sustainability. This chapter sets the stage for the entire book, emphasizing the importance of incorporating various cultural insights to tackle the environmental, social, and economic challenges we face today. It highlights how communities from indigenous groups to urban innovators contribute to building a resilient and sustainable global community, underscoring the interconnectedness of all life and the critical need for collaborative efforts. This opening section motivates readers to actively engage in shaping a future that honors cultural diversity while advancing environmental stewardship.

# Chapter 1:
# Cultural Threads in the Fabric of Sustainability

*Cultural Threads in the Fabric of Sustainability* explores the rich diversity of cultural perspectives and their impact on sustainable practices globally. The chapter highlights how different cultures view and interact with the environment, emphasizing that sustainability is not only a technical issue but a deeply cultural one. By examining the roles of spiritual beliefs, indigenous knowledge, and various environmental philosophies from around the world, it underscores the importance of integrating these diverse viewpoints to create a more effective, equitable, and culturally sensitive approach to sustainability. The chapter argues for a unified yet multifaceted global strategy that respects and incorporates the wisdom and values of all communities, ultimately enriching our collective approach to overcoming environmental challenges.

# Chapter 2:
# The Anthropocene and Our Shared Future

*The Anthropocene and Our Shared Future* discusses the profound impact of human activities on the Earth, marking the beginning of the Anthropocene epoch. It highlights the dual nature of human influence—both destructive and innovative—on the planet's ecosystems and geology. The chapter reflects on the severe alterations caused by industrialization, such as landscape changes, atmospheric carbon increase, and climate change. It calls for a reevaluation of our relationship with nature, proposing a shift toward sustainable coexistence through global solidarity and cultural innovation. By integrating diverse ideologies and cross-disciplinary insights, the chapter emphasizes humanity's capacity for resilience and adaptability. It presents the Anthropocene as an opportunity to redefine progress and success, not just in economic terms, but also through the health of ecosystems and equitable societal structures. This chapter serves as a call to action for embracing environmental stewardship that ensures the well-being of all life on Earth.

# Chapter 3:
# Systems Thinking and Cultural Synthesis

*Systems Thinking and Cultural Synthesis* explores the essential role of systems thinking in understanding and addressing the complexities of ecological sustainability within diverse human societies. It emphasizes how this analytical framework helps recognize patterns, relationships, and the broader consequences of actions within environmental contexts. The chapter argues for cultural synthesis as a way to integrate various cultural insights and sustainability practices, promoting a unified approach to addressing global challenges. By encouraging readers to adopt systems thinking, the chapter aims to foster resilience and a more holistic view of sustainability, highlighting the interconnectedness of human actions and ecological health.

# Chapter 4:
# Embracing Interconnectedness and Interdependence

*Embracing Interconnectedness and Interdependence* explores the critical importance of recognizing the global interconnectedness and interdependence of environmental, societal, and economic factors in achieving sustainability. The chapter argues that our actions have far-reaching impacts across the globe, necessitating a shift away from isolationism towards a framework of mutual coexistence and collaboration. Through examples of cultural sustainability, it illustrates how interconnected approaches can address global challenges like climate change and social inequality, urging a collective effort in stewardship and sustainable development.

# Chapter 5:
# Global Issues, Local Colors

*Global Issues, Local Colors* delves into how different communities across the globe uniquely address sustainability challenges like climate change, reflecting their diverse cultural and geographical backgrounds. It emphasizes the value of local knowledge systems and traditions in crafting effective, culturally resonant responses to environmental crises. Highlighting examples from various cultures, the chapter illustrates the integration of traditional practices with modern technologies to enhance resilience and sustainability. It advocates for the recognition and application of these localized solutions in global sustainability strategies, stressing the importance of embracing the colorful diversity of responses to address universal environmental issues effectively.

# Chapter 6:
# Sustainability and the Language of Culture

*Sustainability and the Language of Culture* explores how cultural narratives shape and drive sustainability efforts by transforming values into actionable behaviors that resonate within various communities. It emphasizes the role of cultural expressions—traditions, heritage, and collective memory—in communicating and implementing sustainability. The chapter highlights the integration of local cultural practices with global sustainability goals, demonstrating that embracing cultural diversity is crucial for developing effective and inclusive environmental strategies. It calls for a deeper understanding and utilization of the cultural context to foster a more sustainable future globally.

# Chapter 7:
# Indigenous Wisdom for Modern Challenges

*Indigenous Wisdom for Modern Challenges* explores the invaluable contributions of Indigenous knowledge to contemporary environmental management and sustainability practices. Highlighting millennia-old techniques for land management, biodiversity enhancement, and sustainable living, the chapter advocates for the integration of these ancient wisdoms into modern policymaking and practices to address pressing global issues like climate change and resource depletion. It underscores the importance of respecting and incorporating the deep ecological insights and cultural values of Indigenous communities to foster a more symbiotic relationship with nature and enhance global sustainability efforts.

# Chapter 8:
# The Role of Religion in Green Living

*The Role of Religion in Green Living* explores how various religious traditions profoundly influence environmental stewardship and sustainability practices. The chapter discusses how faith-based initiatives and moral imperatives within religions drive ecological conservation, emphasizing the integration of scriptural teachings with modern environmental ethics. It highlights how religious communities worldwide are adopting green living practices, demonstrating that spiritual beliefs can effectively bridge the gap between ecological awareness and actionable conservation efforts. This chapter calls for deeper engagement of religious values in promoting sustainable behaviors and policy changes that respect and protect the natural world.

# Chapter 9:
# Art, Music, and the Rhythms of Sustainability

*Art, Music, and the Rhythms of Sustainability* examines how the arts serve as a dynamic force in environmental advocacy, transforming abstract ideas of sustainability into emotive and actionable experiences. This chapter highlights how various forms of art, including music, visual arts, dance, and literature, not only reflect societal values but actively mold public perception and behavior towards environmental issues. Artists and musicians evoke emotional responses and inspire community action, making sustainability deeply resonant on a personal and collective level. By integrating ecological themes into their work, artists contribute to a cultural shift towards more sustainable practices, demonstrating the arts' critical role in fostering a more environmentally conscious society.

# Chapter 10:
# Education as the Seed of Sustainability

*Education as the Seed of Sustainability* emphasizes the pivotal role of education in cultivating a deep-rooted culture of environmental stewardship. It explores the transformative impact of integrating sustainability into educational frameworks across all levels, from creative early learning in schools to continuous professional development in businesses. The chapter advocates for a multicultural curriculum that reflects the diverse global community and prepares students not just as future workforce members, but as proactive guardians of the planet. By embedding sustainable practices and values in education, the chapter argues for establishing a legacy of environmental responsibility that transcends generations, ensuring that each individual recognizes their role in contributing to the Earth's sustainability.

# Chapter 11:
# Green Economics: Weaving Prosperity with Planet

*Green Economics: Weaving Prosperity with Planet* explores the concept of green economics, where ecological sustainability forms the core of economic practices, rather than being an afterthought. It highlights how redefining prosperity to focus on well-being and environmental health leads to innovative business models that align economic success with ecological stewardship. The chapter emphasizes the importance of integrating cultural diversity into these models, showing how a multicultural approach enriches green economics by bringing a variety of sustainable practices and perspectives that can lead to more effective and adaptable solutions. It paints a future where economies thrive by supporting the planet's health and leveraging cultural insights to enhance sustainability practices across global markets.

# Chapter 12:
# Rural Traditions and Innovations

*Rural Traditions and Innovations* examines how rural communities globally blend age-old traditions with modern innovations to foster sustainable development. It highlights the unique position of rural areas in preserving ecological balance and cultural heritage while adapting to contemporary environmental challenges. The chapter underscores the importance of traditional agricultural practices and local knowledge in enhancing biodiversity and food security, showing how these practices are being integrated with modern technology to revitalize rural economies and conserve the environment. Through a focus on the symbiotic relationship between tradition and innovation, the chapter reveals how rural communities are crucial to achieving sustainability, providing lessons in resilience and adaptability that can inform broader global strategies for environmental stewardship.

# Chapter 13:
# The Fabric of Social Sustainability

*The Fabric of Social Sustainability* explores the critical interconnections between social equity, cultural diversity, and environmental sustainability. It emphasizes that true sustainability cannot be achieved without addressing the social foundations of equity and inclusion. The chapter discusses how social justice is intertwined with environmental issues, highlighting the necessity of integrating diverse cultural perspectives and practices into sustainability efforts to ensure no community is left behind. It also delves into various cultural approaches to health, illustrating how traditional knowledge and practices can contribute significantly to the well-being of communities and enhance their resilience. By advocating for a comprehensive approach that considers social, economic, and environmental dimensions, the chapter underscores the importance of building a sustainable society that is equitable, inclusive, and respectful of all cultural identities.

# Chapter 14:
# Policy Patterns for a Sustainable Future

*Policy Patterns for a Sustainable Future* discusses the crucial integration of cultural sensitivity into environmental policy-making and the importance of multilateral cooperation in addressing global sustainability challenges. It emphasizes that effective environmental regulations must consider the cultural, historical, and social contexts of the communities they affect, promoting policies that are not only ecologically sound but also culturally appropriate and just. The chapter highlights the concept of "policy embroidery," describing the intricate weaving of local and global initiatives to create a cohesive and resilient approach to sustainability. This involves harmonizing diverse cultural values with scientific insights and environmental imperatives, ensuring that policies are inclusive and equitable. By fostering multilateral collaborations, these policies aim to unify international efforts, respecting cultural differences while pursuing common environmental goals. The chapter underscores the necessity for continuous adaptation and learning in policy-making to accommodate changing cultural landscapes and ecological realities, advocating for a future where sustainability is deeply embedded in the fabric of global governance and cultural respect.

# Chapter 15:
# Telling the Stories of Climate Change

*Telling the Stories of Climate Change* explores how narratives shape perceptions and actions regarding climate change. It emphasizes the role of media in communicating sustainability, highlighting the need for stories that combine scientific facts with emotive elements to engage broader audiences. The chapter stresses the importance of incorporating diverse voices in climate discourse, particularly from marginalized and indigenous communities, to ensure a comprehensive understanding of climate challenges. By weaving together various narratives from different cultures and perspectives, the chapter advocates for a collective approach to sustainable solutions, underscoring that inclusivity in storytelling is crucial for effective climate action.

# Chapter 16:
# Technology and Tradition: A Delicate Dance

*Technology and Tradition: A Delicate Dance* examines the intersection of modern innovations and ancestral wisdom in promoting global sustainability. This chapter emphasizes the importance of integrating technological advancements with traditional knowledge to create sustainable solutions that respect cultural heritages. It explores how ancient practices like water harvesting and sustainable agriculture are enhanced by modern technologies, creating effective, culturally sensitive applications. The chapter argues for equitable access to green technology, ensuring it benefits all sectors of society, not just the affluent. By honoring and utilizing the deep-rooted wisdom of various cultures, the chapter advocates for a balanced approach to sustainability that bridges the gap between the old and the new.

# Chapter 17:
# Cultural Festivals of Sustainability

*Cultural Festivals of Sustainability* explores how festivals serve as platforms for promoting environmental awareness and sustainable practices. Across the globe, these cultural events merge tradition with ecological stewardship, showcasing the potential for festivals to influence sustainable lifestyle changes and foster a deep respect for nature. The chapter delves into how festivals from various cultures celebrate the Earth through traditional and innovative practices that emphasize conservation. These events not only provide joy and community cohesion but also educate and inspire attendees to adopt sustainable behaviors that contribute to global environmental goals. Through vivid examples of festivals that blend cultural heritage with green initiatives, this chapter highlights the dynamic role of cultural festivals in advancing sustainability.

# Chapter 18:
# Culinary Cultures: A Taste of Sustainability

*Culinary Cultures: A Taste of Sustainability* explores how global food traditions reinforce sustainable practices. This chapter delves into the interplay between local gastronomies and environmental stewardship, illustrating how age-old culinary methods contribute to modern sustainability goals. It highlights the ecological benefits of traditional diets, which often rely on local, seasonal produce, thereby reducing the ecological footprint and enhancing nutritional health. The chapter also discusses how the principles of diversity, moderation, and mindfulness in food preparation and consumption promote a sustainable relationship with the environment. By embracing these culinary practices, societies worldwide can advance towards a more sustainable and resilient future, honoring cultural heritage while fostering global environmental health.

# Chapter 19:
# Fashion and Fabrics: A Material Conscience

*Fashion and Fabrics: A Material Conscience* explores the evolving role of the fashion and textiles industry towards sustainable practices. This chapter underscores the importance of integrating ethical considerations with aesthetic appeal in the production and choice of clothing. It highlights how both traditional practices and innovative technologies are being leveraged to create environmentally friendly apparel that reflects cultural heritage and reduces ecological impact. Emphasizing the significance of conscious consumer choices, it calls for a collective effort to promote sustainable fashion that supports environmental health and cultural expression, urging individuals to influence industry trends by choosing garments that are not only stylish but also sustainably made.

# Chapter 20:
# Activism and Advocacy: Voices for the Earth

*Activism and Advocacy: Voices for the Earth* illuminates the essential role of grassroots movements and individual advocates in driving global sustainability efforts. This chapter celebrates the power of unified voices—from local activists to global networks—pushing for environmental preservation and policy change. It delves into how storytelling serves as a crucial tool, turning personal and community experiences into compelling calls for action that resonate worldwide. Highlighting the symbiotic relationship between local initiatives and global impacts, the text encourages readers to recognize their own potential as part of this dynamic wave of change, urging them to raise their voices for the health of the planet.

# Chapter 21:
# The Symphony of Sustainable Communities

*The Symphony of Sustainable Communities* explores the intricate balance of sustainable living within communities, likening it to a well-conducted orchestra where diverse elements harmonize to create a resilient and thriving environment. This chapter delves into the strategies for aligning economic, environmental, and social interests to promote collective well-being. It highlights how sustainable communities, like music ensembles, require each member to contribute their unique strengths in concert with others. The text encourages readers to view their community interactions as parts of a greater whole, emphasizing cooperation, shared values, and mutual support as the foundation for sustainable success.

# Chapter 22:
# The Unfinished Quilt

*The Unfinished Quilt* serves as a reflective conclusion to the journey through sustainable practices, likening the ongoing global efforts to an expansive, yet incomplete quilt. This chapter revisits the myriad cultural, technological, and ecological insights previously explored, emphasizing that sustainability is a dynamic, evolving process rather than a finalized state. It acknowledges the essential contributions of diverse cultures, the integration of traditional wisdom with modern innovation, and the crucial role of systems thinking in weaving a coherent, sustainable future.

Through this quilt metaphor, the chapter inspires continued dedication to sustainability, urging readers to see their actions as vital stitches in the ever-expanding tapestry of human and ecological harmony. It calls for an embrace of the quilt's unfinished nature, symbolizing our unending responsibility to nurture and repair our world, ensuring that every new stitch contributes to a resilient, equitable, and thriving planet.

# Appendix:
# Cultural Case Studies in Sustainability

In *Appendix: Cultural Case Studies in Sustainability*, we delve into a series of cultural case studies that underscore the diverse, innovative approaches to sustainability practiced by communities across the globe. From the ancient terraced fields of the Philippines to the frozen tundra of the Arctic, these examples demonstrate how local knowledge and practices can contribute significantly to global sustainability efforts, offering valuable insights into harmonizing cultural heritage with environmental stewardship.

- The Ifugao Rice Terraces, Philippines: These ancient terraces are not just agricultural feats but also a sophisticated system of water management and soil conservation that supports both biodiversity and the local community.

- The Waorani People, Ecuador: Guardians of the Amazon, the Waorani utilize extensive knowledge of their forest ecosystem to manage it sustainably while fighting to protect their land from oil extraction.

- Inuit Ice Fishing, Arctic: Demonstrating sustainable hunting practices, Inuit communities use ice fishing techniques that ensure fish populations remain stable, supporting their subsistence lifestyles without depleting resources.

- Sámi Reindeer Herding, Scandinavia: This traditional practice is an example of sustainable animal husbandry where herding techniques are in harmony with the environment, ensuring the health of both reindeer populations and the local ecosystem.

- The Mbuti Pygmies, Democratic Republic of the Congo: Inhabitants of the Ituri Forest, the Mbuti practice a form of sustainable hunting and gathering that has minimal impact on their environment, maintaining biodiversity.

- The Tuareg People, Sahara Desert: Nomadic lifestyles of the Tuareg involve moving herds across vast areas to prevent overgrazing, using traditional knowledge to survive in one of the harshest climates on earth.

- The Qanat System, Iran: An ancient ingenuity in sustainable water management, the Qanat system channels underground water to communities without the use of pumping technology, preserving water resources in arid regions.

- The Hunza Community, Pakistan: Known for their longevity and health, the Hunza's use of terraced farming and natural remedies exemplifies sustainable living practices that are integrated into cultural traditions.

- Brokpa Community, Bhutan: Living in one of the most remote areas of the Himalayas, the Brokpa people practice agro-pastoralism, meticulously managing their natural resources in a way that supports both their community and the surrounding environment.

- Pamiri Houses, Tajikistan: These traditional homes built from local materials not only utilize sustainable construction practices but also incorporate passive solar design to maintain temperature, showcasing a harmonious balance between human habitation and environmental preservation.

These case studies are not only testaments to the resilience and ingenuity of various cultures in the face of environmental challenges but also highlight the importance of preserving these practices. They remind us that sustainable solutions often lie in the wisdom of traditional practices that have evolved in close harmony with nature. As we look to the future, integrating this knowledge into broader environmental strategies becomes crucial for the sustainability of our planet.

# A Sustainable Civilization

### Unraveling the Threads of Change

## CHAPTER SYNOPSIS

### James Fountain

Our Changing World

# Book 2: A Sustainable Civilization

*Unraveling the Threads of Change*

# A Sustainable Civilization: Unraveling the Threads of Change

Book 2 of the Our Changing World series, A Sustainable Civilization: Unraveling the Threads of Change, continues the exploration begun in Threads of Green. It embarks on a comprehensive exploration of the sustainability movement, tracing its historical roots, navigating its current challenges, and envisioning its future trajectory. This volume highlights the pivotal contributions of diverse cultures and integrates ancient wisdom with contemporary principles. By doing so, it addresses the multifaceted challenges posed by industrialization, population growth, and rapid technological advancements. The book meticulously weaves these elements together, illustrating how traditional knowledge and modern innovations can collectively forge a sustainable path forward.

The narrative extends into a detailed examination of specific environmental issues such as climate change, renewable energy, water conservation, biodiversity, and regenerative agriculture, which are meticulously linked back to the cultural and global perspectives discussed earlier. This linkage underscores the interconnectedness of human actions and environmental impacts, reinforcing the themes of cultural diversity and global unity in addressing these pressing challenges. The discussion on emerging issues like the role of technology in sustainability is particularly poignant, reflecting the dynamic interplay between technological advances and sustainable practices.

Central to the narrative are the influential figures and landmark events that have significantly molded the sustainability discourse, offering readers an inspirational glimpse into the evolving ethical landscape that seeks to redefine humanity's bond with nature. Through engaging stories and pivotal examples, the book delves into the critical battles against climate change and biodiversity loss, underscoring the indispensable role of Indigenous knowledge in weaving a resilient and sustainable future. These discussions are pivotal in illustrating how historical and current efforts in various fields of environmental concern are crucial for the development of sustainable solutions.

By presenting a vision of a world where humanity and nature exist in harmonious coexistence, the book aspires to inspire a collective movement towards sustainable living. It concludes with a powerful vision for a sustainable future, portraying a scenario where the lessons of the past and the innovations of the present converge to create a sustainable civilization. This vision emphasizes the importance of unity, innovation, and respect for the natural world in shaping the future we aspire to leave for generations to come, culminating in a call to action for readers to contribute to this ongoing and vital global effort.

# Preface

The preface of the *Our Changing World* series delves into the critical understanding of our impact on the planet, urging an immediate and reflective engagement with the ecological, cultural, and governance threads that shape our existence. This second book in the series emphasizes the importance of recognizing and acting upon the continuous changes affecting our world. By weaving together diverse perspectives on ecological thought, cultural sensitivity, and sustainable governance, it sets a narrative stage that calls for active participation in shaping a sustainable future. The book serves as a continuation of the themes introduced in "Threads of Green," focusing on deepening the exploration of cultural diversity, sustainability, and ethical governance, which are pivotal for addressing the challenges of our dynamic planet.

# Chapter 1:
# The Imperative for Change: Embarking on Sustainability

*The Imperative for Change: Embarking on Sustainability* outlines the urgent need for a shift toward sustainability amid growing environmental uncertainties and degradation, emphasizing that our current actions will significantly impact future generations. The chapter argues for a collaborative and scientifically informed approach across ecological, economic, and societal domains, urging the embrace of innovative solutions and transformative changes to support the resilience of the natural world and human spirit. It presents sustainability not just as an option but as an imperative—vital for the long-term survival of both civilization and ecosystems. The narrative stresses the importance of balancing current needs with the future's, underlining sustainability as a matter of social justice and cultural preservation, ensuring equitable access to resources for all communities, particularly those most affected by environmental issues. The chapter concludes by calling for a holistic view that integrates economic growth with environmental conservation and social equity, setting the stage for comprehensive systemic changes needed to achieve a sustainable future.

# Chapter 2:
# Lessons from History:
# Tracing Our Sustainable Roots

*Lessons from History: Tracing Our Sustainable Roots* explores the deep-rooted history of sustainable practices across various ancient civilizations and indigenous cultures. It highlights how these societies were inherently connected to their environments, employing cyclical resource use, biodiversity preservation, and dynamic equilibrium to ensure their survival and prosperity. These practices were not only sustainable but also respectful of the natural world, suggesting a form of environmental stewardship that modern societies can learn from. The chapter argues that while modern innovations are essential, there is immense value in rediscovering and integrating these ancient sustainable practices into current efforts to address environmental challenges. By examining historical models of sustainability, the chapter provides a blueprint for how we can enhance contemporary approaches to sustainability, making them more holistic and effective.

# Chapter 3:
# A World in Flux: Confronting Climate Realities

*A World in Flux: Confronting Climate Realities* tackles the pressing issues of climate change, examining its scientific underpinnings and the significant impacts it has on our ecosystems and societies. It emphasizes the dynamic nature of our planet's climate, detailing how changes affect biodiversity, cultural heritage, and global economic structures. This chapter calls for an urgent and innovative approach to climate mitigation and adaptation, advocating for integrated solutions that encompass economic, ecological, and technological perspectives. It also underscores the importance of community engagement and comprehensive policy reform in developing effective strategies to enhance societal resilience. Through scientific inquiry and a collaborative approach, the chapter seeks to equip readers with the knowledge and tools necessary to navigate the complexities of our changing climate and to work towards a sustainable and resilient future.

# Chapter 4:
# The Renewable Revolution: Harnessing Energy

*The Renewable Revolution: Harnessing Energy* delves into the transformative shift towards renewable energy sources such as solar, wind, and hydro, which are increasingly central to our global energy infrastructure. This chapter outlines the technological advancements and economic shifts that facilitate this crucial transition, highlighting how these energy sources not only contribute to reducing carbon emissions but also play a vital role in shaping a sustainable future. The discussion extends to the broader implications of this shift, including changes in societal norms, economic structures, and the formulation of global energy policies. The chapter advocates for a holistic approach, suggesting that the integration of renewable energy into mainstream society involves rethinking energy consumption patterns and embracing sustainable solutions at both local and global levels.

# Chapter 5:
# The Waters of Life: Quenching a Thirsty World

*The Waters of Life: Quenching a Thirsty World* underscores the urgent need to preserve and enhance the world's water resources amid escalating climate crises and population growth. It explores the delicate balance required to manage water sustainably, addressing the challenges posed by scarcity, urbanization, and the extensive energy requirements of water purification processes. This chapter highlights collaborative efforts across communities, governments, and businesses, emphasizing the universal importance of water stewardship. It presents a comprehensive view of the strategies needed to manage and conserve water effectively, ensuring its availability for future generations while discussing the roles of various stakeholders in fostering a sustainable approach to water management.

# Chapter 6:
# Oceanic Heritage: Guarding the Marine Commons

*Oceanic Heritage: Guarding the Marine Commons* delves into the vital importance of the world's oceans as immense reservoirs of biodiversity and crucial regulators of the global climate. It emphasizes the oceans' indispensable role in sustaining life on Earth through their life-sustaining functions and explores the severe threats they face, such as overfishing, pollution, and acidification. The chapter advocates for robust conservation strategies and effective governance to safeguard marine environments, underscoring the intrinsic link between the oceans' health and our own survival. It highlights the transformative potential of Marine Protected Areas (MPAs) and innovative policy frameworks that embody the ethos of sustainability, offering hope and actionable solutions to preserve the marine commons for future generations.

# Chapter 7:
# Waste Not:
# The Circular Economy's March Forward

*Waste Not: The Circular Economy's March Forward* discusses shifting from a linear "take-make-dispose" model to a circular economy that minimizes waste and reuses materials. This approach not only lessens environmental impact but also bolsters economic stability and social cohesion. The chapter highlights innovations in recycling and sustainable resource management that support this transition. It advocates for a future where economic growth is sustainable, advocating for systemic changes that include designing products for longevity and promoting resource efficiency to achieve environmental and economic benefits.

# Chapter 8:
# Biodiversity's Beacon:
# Safeguarding Our Natural Heritage

*Biodiversity's Beacon: Safeguarding Our Natural Heritage* emphasizes the critical importance of biodiversity, describing it as a complex network essential for ecological and human health. The chapter examines the severe impacts of biodiversity loss and explores strategies for species protection and habitat preservation. Highlighting conservation efforts as both an ethical obligation and a necessity for survival, it calls for unified action to protect our natural heritage through sustainable practices and policies that support biodiversity integrity.

# Chapter 9:
# Cultivating Tomorrow:
# Seeds of Agricultural Innovation

*Cultivating Tomorrow: Seeds of Agricultural Innovation* explores the transformative shift in agriculture towards sustainable practices that meld traditional knowledge with modern science. It highlights the necessity for agriculture to adapt through precision farming, biotechnology, and soil health enhancement amid challenges like climate change and population growth. The chapter champions sustainable agriculture as crucial for maintaining food security and ecological health, promoting techniques that reduce environmental impact and enhance resilience. By fostering innovations in sustainable farming, the chapter envisions agriculture not only as a means to produce food but as a stewardship of ecological balance.

# Chapter 10:
# New Horizons:
# Technological Frontiers in Sustainability

*New Horizons: Technological Frontiers in Sustainability* discusses the pivotal role of emerging technologies in advancing sustainability goals. The chapter highlights innovations across various fields like renewable energy storage, biodegradable materials, precision agriculture, and smart urban planning, emphasizing how these technologies are reshaping our approach to environmental conservation. It underscores the transformative impact of integrating ethical considerations with technological advancements, ensuring that progress in sustainability is balanced and beneficial for all. This exploration confirms technology's crucial role in both mitigating environmental impact and enhancing the symbiotic relationship between human progress and ecological preservation.

# Chapter 11:
# Weaving the Threads:
# Co-Creating Tomorrow's Sustainability

*Weaving the Threads: Co-Creating Tomorrow's Sustainability* serves as the concluding chapter of the book, articulating an integrated and collective approach to achieving sustainability. This chapter emphasizes the synergy between technology, ethical practices, and ecological insights, advocating for a unified strategy that leverages diverse fields to address environmental challenges effectively. Highlighting the necessity of collaboration across all sectors of society, it underscores the role of renewable technologies, biodiversity protection, and systemic changes in waste management and energy consumption. It portrays sustainability not just as a series of actions but as a dynamic, co-creative journey that requires shared stewardship and proactive involvement to build a resilient and sustainable future for all.

James Fountain

# A Planet in Balance

## Exploring the Socio-Cultural Landscape of Sustainability

## CHAPTER SYNOPSIS

### James Fountain

Our Changing World

# Book 3: A Planet in Balance
*Exploring the Socio–Cultural Landscape of Sustainability*

# A Planet in Balance: Exploring the Socio-Cultural Landscape of Sustainability

*A Planet in Balance: Exploring the Socio-Cultural Landscape of Sustainability*, the third book in the *Our Changing World* series, deepens the exploration of sustainability by focusing on its ethical and socio-cultural dimensions. It effectively connects the ethical imperatives and cultural insights to the practical sustainability challenges discussed in the previous books. By examining key themes such as environmental justice, social equity, and human rights, the book underscores the necessity of an ethical, inclusive approach to sustainability. This approach highlights how cultural understanding and ethical considerations are integral to effective global stewardship and policymaking.

The book provides a thorough examination of the intricate relationships between sustainability, ethics, and multicultural perspectives. It delves into the ethical bases that support the pillars of environmental, social, and economic sustainability, presenting a nuanced discussion on how these ethical considerations are essential for a deep understanding and advancement of sustainability goals. The narrative explores how individual values, cultural backgrounds, and societal frameworks critically shape our perceptions and engagements with sustainability, demonstrating the complex interplay between personal beliefs and global challenges.

Central to the book's discourse is the vital role that education, art, and storytelling play in enhancing sustainability awareness and motivating societal action. These elements are portrayed as powerful tools for bridging the gap between knowledge and action, effectively engaging diverse audiences and fostering a deeper connection to sustainability issues. The narrative emphasizes how these tools can illuminate the impacts of sustainability on everyday life and inspire positive change through compelling narratives and artistic expression.

The discussion extends into the realm of global governance and international cooperation, scrutinizing the mechanisms of policy development and the execution of international environmental agreements. This examination reveals the complexities and challenges of global collaboration, while also highlighting successful strategies that have led to meaningful progress in sustainability efforts.

By providing an in-depth analysis of how ethics, culture, and governance intersect with sustainability, A Planet in Balance offers a comprehensive perspective on the socio-cultural dimensions of sustainability. The book encourages readers to reflect on their roles within this global framework and inspires collective efforts towards achieving a sustainable and balanced planet, making it an essential read for anyone interested in the deeper cultural and ethical aspects of sustainability.

# Preface

The Preface of *A Planet in Balance* frames sustainability as not just an environmental or economic challenge, but as a profound ethical and cultural issue. This chapter invites a broad spectrum of stakeholders, including academics, professionals, and activists, to engage with the ethical underpinnings and cultural narratives that shape our approach to sustainability. It argues for a deeper integration of ethical values in our sustainability efforts, emphasizing that our environmental crisis is tightly linked to our moral choices and societal values. By exploring eco-ethics, the chapter sets the stage for viewing sustainability through a lens of ethical stewardship and cultural inclusivity, laying a foundational vision for a sustainable and equitable future.

# Chapter 1:
# Ethical Foundations of Sustainability

*Ethical Foundations of Sustainability* outlines the necessity of integrating ethical considerations into sustainability efforts. It emphasizes that sustainability transcends environmental and economic concerns, incorporating profound ethical challenges that dictate our stewardship of the planet. The chapter introduces key ethical theories—utilitarianism, deontology, virtue ethics, and care ethics—that guide sustainable practices. It discusses how these theories influence decision-making processes in sustainability, highlighting the importance of considering the impacts on both current and future generations. The narrative interweaves social, cultural, and environmental justice, advocating for a holistic approach that respects and nurtures the interconnectedness of all life forms and ecosystems.

# Chapter 2:
# Cultural Narratives and Socio-Cultural Understanding in Sustainability

*Cultural Narratives and Socio-Cultural Understanding in Sustainability* explores the vital role of cultural narratives in shaping sustainable practices. It emphasizes that cultural diversity and socio-cultural understanding are crucial for effective sustainability strategies. The chapter discusses how various cultural perspectives influence environmental interactions and sustainability efforts, highlighting traditional and Indigenous practices that offer sustainable solutions. It argues against a one-size-fits-all approach, advocating for integrating diverse cultural insights to foster resilience and innovation in sustainability. The narrative showcases how respecting and incorporating these diverse cultural dimensions can lead to more comprehensive and effective sustainability outcomes globally.

# Chapter 3:
# Sustainability Perspectives on Environmental Justice

*Sustainability Perspectives on Environmental Justice* explores the critical link between social equity and environmental sustainability. It discusses how true sustainability cannot be achieved without addressing the injustices that affect marginalized communities. The chapter highlights the role of the environmental justice movement in advocating for fair distribution of environmental benefits and burdens, emphasizing the need for systemic change to ensure fairness in environmental policy. Through examples of global activism, it illustrates the universal nature of environmental justice issues and the imperative to integrate social equity into sustainability efforts, ultimately advocating for a more equitable, ethical, and resilient world.

# Chapter 4:
# Social Equity and Sustainability

*Social Equity and Sustainability* explores how social equity is essential for true sustainability. It discusses the importance of creating fair and inclusive systems that allow all individuals to thrive and reach their potential, emphasizing that sustainability must balance human and ecological systems. The chapter examines the impact of income inequality, environmental racism, and the need for inclusive strategies that address these disparities. It advocates for integrating social equity into sustainability by ensuring equitable access to resources, inclusive urban planning, and fair transitions to green jobs. Overall, the chapter highlights the interconnectedness of social equity and environmental integrity in achieving a sustainable future.

# Chapter 5:
# Examining Critical Social Justice Issues in Sustainability

*Examining Critical Social Justice Issues in Sustainability* delves into the crucial role of justice within the sustainability narrative, underscoring its importance in maintaining ecological balance and fostering social harmony. The chapter outlines justice's multifaceted function in promoting equity, ensuring fair resource distribution, and valuing natural preservation for both present and future generations. Through an examination of historical contexts and root causes, the text explores the transition from exploitation to ethical stewardship, highlighting how past actions influence current environmental challenges. This historical perspective helps to trace the evolution of activism and policy, offering insights into the systemic failures and ongoing inequalities that exacerbate resource scarcity, climate vulnerabilities, and energy consumption inequities among marginalized groups. The chapter invites readers to critically assess these dynamics, reflecting on individual and collective roles within the ongoing saga of social and ecological justice.

# Chapter 6:
# Human Rights in Sustainability

*Human Rights in Sustainability* discusses the vital integration of human rights within sustainability practices. It asserts that every individual's right to a life of dignity inherently includes access to a healthy environment. This chapter emphasizes that the pursuit of sustainability extends beyond environmental health to ensuring equitable access to natural resources, protection from environmental harm, and inclusive participation in decision-making processes. Recognizing environmental degradation as a violation of human rights, the chapter argues for sustainable solutions that promote mutual flourishing and collective stewardship, shifting the narrative from exploitation to one of coexistence and respect for all life forms.

# Chapter 7:
# Reimagining Cities for a Sustainable Tomorrow

*Reimagining Cities for a Sustainable Tomorrow* explores the transformation of urban environments into sustainable cities through the integration of social, economic, and ecological considerations. It highlights how sustainable cities are not just about green architecture but involve holistic development strategies that ensure health, equity, and resilience. The chapter discusses the use of green technologies, ethical planning, and community-led initiatives as central to turning cities into hubs of sustainable innovation. These efforts are framed as opportunities to improve the quality of life while ensuring environmental stewardship, emphasizing the critical role of current urban planning and civic engagement in shaping sustainable future metropolises.

# Chapter 8:
# The Global Collaboration for a Sustainable Future

*The Global Collaboration for a Sustainable Future* highlights the critical need for global collaboration in pursuing a sustainable future. It underscores the importance of global governance and international cooperation, emphasizing how these frameworks facilitate the ethical management of environmental issues on a worldwide scale. The chapter also details the significant role of activism in driving change, noting its capacity to mobilize global communities towards unified actions. By weaving together different cultural perspectives and political intricacies, the chapter advocates for a cohesive approach to sustainability, proposing that the combined efforts of governments, corporations, and citizens are essential to address the complex challenges of global sustainability effectively.

# Chapter 9:
# Balancing Economic Systems and Business Ethics in Sustainability

*Balancing Economic Systems and Business Ethics in Sustainability* explores the crucial balance between economic systems and sustainability, emphasizing the integration of business ethics with ecological responsibility. It critiques capitalism's ecological footprint and advocates for alternative economic models that combine prosperity with environmental integrity. The chapter delves into sustainable business practices and the growing importance of Corporate Social Responsibility (CSR) and Environmental, Social, and Governance (ESG) metrics in reshaping corporate influence towards nurturing both humanity and the planet. It calls for innovative changes not only in technology but also in ideological approaches, promoting a market system where ethical and ecological considerations are inherently aligned with business operations.

# Chapter 10:
# Ethical Sustainability Dialogues in Media, Faith, Education, and Science

*Ethical Sustainability Dialogues in Media, Faith, Education, and Science* delves into the crucial roles that media, faith, education, and science play in fostering ethical sustainability dialogues, which are essential for nurturing a conscious and sustainable society. The media influences public perception and action by shaping societal values and priorities through narratives that emphasize the urgency of stewardship. Various faith traditions contribute insights and call for eco-ethical actions that align with their teachings on stewardship and the sanctity of nature. Education enhances eco-literacy and cultural competence, essential for understanding and addressing global sustainability challenges. Science supports these efforts by providing a factual basis for sustainability practices and by integrating ethical considerations into research, ensuring that scientific advancements contribute positively to sustainability goals. Together, these sectors create a robust dialogue that encourages the public to engage in conscientious stewardship of the planet.

# Chapter 11:
# Cultivating a Future of Collective Responsibility and Sustainable Innovation

*Cultivating a Future of Collective Responsibility and Sustainable Innovation* delves into the interplay between innovation and collective responsibility in achieving sustainability. It argues for integrating ethical considerations into innovation to ensure technological advancements support a sustainable future for all. Highlighting emerging technologies like renewable energy and precision agriculture, the chapter emphasizes their transformative potential while cautioning against unintended ecological impacts. The chapter advocates for a paradigm shift where sustainability is integral to societal design and individual actions, calling for policies that align technological progress with environmental stewardship and social equity.

# Conclusion: Integrating a Balanced Planet into Global Sustainability

The conclusion, *Integrating a Balanced Planet into Global Sustainability*, underscores the imperative of integrating ethical, social, and technological dimensions to foster a sustainable future. It advocates for a paradigm shift that respects cultural narratives, champions environmental justice, and upholds human dignity. By merging traditional knowledge with modern science, we can develop innovative, culturally resonant solutions that address global sustainability effectively. The chapter calls for a collective awakening to our interconnected fate, emphasizing solidarity and ethical consciousness as essential for achieving a balanced planet where life flourishes in harmony with nature.

# Appendix A: Ethical Frameworks and Sustainability

*Appendix A: Ethical Frameworks and Sustainability* highlights the critical integration of moral reasoning with scientific innovation in steering towards a sustainable future. It explores various ethical theories, including virtue ethics, rights-based approaches, and the ethics of care, emphasizing their role in promoting environmental integrity and social justice. These frameworks encourage practices that respect the intrinsic value of the natural world and humanity's role within it, guiding actionable strategies across diverse global contexts. The discussion underlines the necessity of empathy, mutual respect, and ethical consciousness in shaping sustainability policies and practices, ensuring they are culturally resonant and ecologically viable.

# Appendix B: Cultural Studies and Environmental Justice Resources

*Appendix B: Cultural Studies and Environmental Justice Resources* underscores the vital intersection of cultural studies and environmental justice in shaping sustainable futures. It stresses the importance of understanding cultural dynamics and integrating indigenous and local knowledge into environmental policies for more effective and respectful sustainability practices. The appendix highlights the role of environmental justice in ensuring equitable treatment in environmental policy and practice, emphasizing the need for all voices, especially marginalized ones, to be heard in decision-making processes. It also explores eco-criticism, which examines cultural products to understand environmental attitudes. The text advocates for embedding principles of environmental equity and cultural respect in education to prepare future generations for holistic and just approaches to sustainability challenges.

James Fountain

# The Responsibility Renaissance

Business as a Catalyst for Environmental and Social Ethics

## CHAPTER SYNOPSIS

### James Fountain

James Fountain

## Our Changing World

# Book 4: The Responsibility Renaissance

*Business as a Catalyst for Environmental and Social Ethics*

# The Responsibility Renaissance: Business as a Catalyst for Environmental and Social Ethics

The Responsibility Renaissance: Business as a Catalyst for Environmental and Social Ethics, the final book in the Our Changing World series, shifts the focus to the business world, illustrating how sustainability transcends cultural and ethical domains to become a critical business imperative. The book links the concepts of corporate responsibility and ecological economics with the broader themes of cultural diversity, ethical sustainability, and global interconnectedness explored in earlier volumes. It extensively covers sustainable business strategies, ecological economics, corporate sustainability, sustainable supply chain management, sustainable innovation, consumer behavior, corporate social responsibility (CSR), and the integration of Environmental, Social, and Governance (ESG) factors. This volume demonstrates how businesses can—and must—play a pivotal role in driving sustainable change, influenced by the cultural, ethical, and socio-economic contexts previously discussed.

The Responsibility Renaissance marks the culmination of insights within the Our Changing World series, emphasizing how businesses are uniquely positioned to drive sustainability and ethical practices within the global marketplace. The book advocates for a significant paradigm shift, urging businesses to adopt regenerative, waste-minimizing, and resource-efficient practices. Through insightful case studies and actionable insights, it showcases the indispensable role of data and analytics in aligning business operations with environmental sustainability goals.

Highlighting that sustainability efforts are more than ethical obligations—they are strategic differentiators—the book provides a variety of strategies for businesses to embed eco-friendly practices throughout their operations. This includes leveraging cutting-edge technologies for sustainable supply chain management and innovating in product and service development. The pivotal role of leadership in driving the sustainability agenda is underscored, emphasizing visionary leaders' capacity to integrate sustainability with innovation, ethical marketing, and impactful community engagement.

Stressing compliance with Environmental, Social, and Governance (ESG) standards as foundational to sustainable business development, The Responsibility Renaissance serves as an essential blueprint for businesses at any stage of their sustainability journey. By weaving environmental and social ethics into the fabric of business practices through theoretical exploration and practical examples, it charts a path for businesses striving to be at the forefront of corporate responsibility and sustainability. This book is an indispensable resource for leaders and entrepreneurs aiming to enact meaningful change and position their enterprises as leaders in the sustainable business arena.

# Preface

The preface of this book emphasizes the significant role of businesses in shaping a sustainable future. It asserts that businesses today are not just about profit maximization but should also act as stewards of the planet and advocates for equitable societies. The book aims to inspire practical action, showing that integrating environmental and social ethics into business practices is both necessary and feasible. Highlighting stories from innovators and visionaries, it challenges conventional business paradigms and offers new perspectives on sustainability that go beyond technological innovation to include cultural and ethical dimensions. This preface sets the stage for a transformative journey, inviting readers to engage deeply with the upcoming discussions on redefining business excellence in harmony with ecological and social responsibilities.

# Chapter 1:
# Navigating the Paradigm Shift in Business and Sustainability

*Navigating the Paradigm Shift in Business and Sustainability* discusses the urgent need for businesses to redefine their role in light of escalating environmental and social challenges. It highlights a transformative shift from traditional profit-centric models to strategies that prioritize ecological integrity and social equity. This change is driven by a recognition that sustainable practices are not just ethical but are also fundamental to long-term business success. The chapter elaborates on how industries are integrating sustainability into their core strategies, balancing profit-making with planetary stewardship, and redefining value creation through innovations in product design, supply chain management, and energy efficiency. It underscores the importance of leadership, creativity, and a commitment to ethical standards as businesses adapt to this new era. The narrative invites businesses, academics, and practitioners to collaborate on sustainable advances, urging a comprehensive reevaluation of corporate practices and strategies to align with environmental and societal imperatives.

# Chapter 2:
# Innovating for the Environment: Business Strategies for the Twenty-First Century

*Innovating for the Environment: Business Strategies for the Twenty-First Century* focuses on how businesses are becoming the architects of a sustainable future through innovative strategies that prioritize ecological stewardship and social responsibility. It emphasizes a transformation from traditional profit-centric models to a more holistic approach that integrates eco-friendly product design, material sourcing, and green technologies. The chapter explores the concept of the circular economy, where waste is minimized, and resources are reused, repaired, and recycled, creating a closed-loop system that benefits both the planet and business sustainability. Furthermore, it discusses the importance of redefining innovation to include eco-friendly and socially conscious approaches, highlighting how these can lead to new market opportunities, enhance brand reputation, and improve operational efficiencies. The narrative urges businesses to rethink their models and practices to align with sustainable principles, driving a significant shift towards a more resilient and regenerative global culture.

# Chapter 3:
# Navigating the Ethical Quandary: Moral Philosophies in Business

*Navigating the Ethical Quandary: Moral Philosophies in Business* examines the integration of moral philosophies into business practices to balance profit with social and environmental responsibilities. It discusses how ethical considerations, once optional, are now essential for corporate longevity and sustainability. The chapter encourages businesses to adopt environmental ethics, viewing themselves as stewards of both profits and the planet. This shift towards ethical business involves rethinking roles and operations, emphasizing a triple bottom line approach that measures success in terms of impact on profit, people, and the planet. Ultimately, it calls for leadership that embeds these values throughout the organization, fostering a culture where ethical practices are the norm.

# Chapter 4:
# The Power of Diversity:
# Cultural Perspectives on Sustainability

*The Power of Diversity: Cultural Perspectives on Sustainability* explores how embracing cultural diversity can enhance sustainability in business. The chapter argues that diversity is not just a source of strength but a critical element for innovation and resilience. By integrating varied cultural viewpoints into their sustainability practices, businesses can access a broader range of knowledge and ideas, leading to more effective and inclusive strategies. This approach helps companies become leaders in sustainability, crafting practices that are respectful of cultural differences and promoting equity alongside environmental stewardship. Overall, the chapter highlights the importance of incorporating diverse cultural insights to drive transformative changes in business towards greater ecological and social responsibility.

# Chapter 5:
# The Local vs. Global Paradigm in Ethical Business

*The Local vs. Global Paradigm in Ethical* Business examines the tension businesses face in balancing local commitments with global responsibilities. It explores how businesses can act as custodians of local culture and traditions while contributing to global sustainability goals. The chapter discusses inclusive business models that integrate local communities into global value chains, thereby fostering local economic growth and sustainable practices simultaneously. It highlights the importance of respecting local customs, engaging in active dialogue with community leaders, and integrating local needs into global strategies. The chapter argues for a nuanced understanding of socio-ecological systems and a deep commitment to community well-being as essential for businesses aiming to thrive ethically in a globalized marketplace.

# Chapter 6:
# The Human Rights Dimension
# of Sustainable Business

*The Human Rights Dimension of Sustainable Business* underscores that respecting human rights is fundamental to ethical corporate practices and crucial for sustainability. It articulates that businesses must regard human rights not as optional but as integral to their operations, impacting employees, suppliers, and communities. This chapter emphasizes how ethical businesses serve as protectors of dignity and equality by ensuring fair wages, safe working conditions, and opposing forced labor. It highlights the necessity for businesses to embed human rights into their ethics through rigorous due diligence, stakeholder engagement, and comprehensive education across the organization. The chapter concludes that the flourishing of humanity is intertwined with the health of our planet, and businesses that recognize and act upon this connection lead in fostering a sustainable future.

# Chapter 7:
# The Sustainable Supply Chain:
# From Raw Materials to End Users

*The Sustainable Supply Chain: From Raw Materials to End Users* discusses the critical importance of sustainable supply chains, from the extraction of raw materials to delivering products to end users. It emphasizes that each step of the supply chain has a significant impact on both the environment and communities, urging businesses to adopt practices that honor and protect both. The chapter highlights the role of ethical sourcing and innovations in logistics that reduce environmental footprints, such as smarter planning and cleaner technologies. Moreover, it explores the challenges businesses face in implementing sustainable supply chain practices, including cost implications, lack of transparency, and the need for industry-wide standardization and collaboration.

# Chapter 8:
# The Technological Revolution and Sustainable Practice

*The Technological Revolution and Sustainable Practice* explores the crucial role of technological innovation in enhancing sustainable business practices. It discusses how digital transformation aids in achieving environmental stewardship by improving supply chain logistics, enhancing carbon accounting systems, and enabling more transparent and accountable corporate practices through real-time reporting. The chapter underscores technologies such as IoT, AI, blockchain, and data analytics as pivotal in optimizing resource use, ensuring ethical sourcing, reducing waste, and promoting energy efficiency. These technologies not only facilitate a deeper integration of sustainability into core business strategies but also drive significant environmental and social improvements, illustrating a synergistic relationship between technological advancement and sustainable development.

# Chapter 9:
# Ethics of Artificial Intelligence and Data in Sustainability

*Ethics of Artificial Intelligence and Data in Sustainability* explores the ethical challenges and considerations of using artificial intelligence (AI) and data analytics in sustainability. It addresses the critical balance between leveraging advanced technologies to enhance sustainable practices and ensuring ethical standards are met, particularly regarding data integrity and privacy. The chapter discusses the potential of AI to significantly improve resource management, predict environmental trends, and optimize supply chains, while also highlighting risks such as bias, privacy breaches, and the perpetuation of inequalities. It emphasizes the need for robust ethical frameworks and regulations to guide the responsible use of AI and data, ensuring that technological advancements in sustainability are both effective and morally sound. The chapter calls for a proactive approach to integrating ethics into technological practices, advocating for transparency, inclusivity, and accountability in the deployment of AI in the field of sustainability.

# Chapter 10:
# Consumerism and Responsibility: Redirecting Behaviors and Demands

*Consumerism and Responsibility: Redirecting Behaviors and Demands* delves into the pivotal role consumer choices play in driving sustainable practices and shaping market dynamics, under the theme "Consumerism and Responsibility: Redirecting Behaviors and Demands". It explores how individual purchasing decisions, influenced by personal ethics and the availability of sustainable choices, impact broader environmental and social outcomes. The chapter emphasizes the need for businesses to educate and engage consumers in making responsible choices through transparent and compelling marketing that aligns with consumer values and beliefs. It suggests that fostering an environment where sustainable choices are intuitive and accessible can lead to a symbiotic relationship between consumer behavior and ethical business practices, ultimately driving a cycle of sustainability in the market. The narrative envisions a future where both supply and demand are oriented toward mutual reinforcement of sustainable practices, with consumer responsibility being integral to achieving broader sustainable development goals.

# Chapter 11:
# Societal Expectations and Business Responses

*Societal Expectations and Business Responses* explores the dynamic relationship between evolving societal demands for sustainability and the strategic responses of businesses. The chapter outlines how modern companies are no longer mere economic entities but pivotal actors within a broader societal and environmental framework, driven by stakeholders who demand ethical practices and sustainability. Businesses are shown to actively integrate these societal expectations into their operations, not just reacting passively but proactively shaping their corporate identities to embrace and promote responsibility and sustainability. Through strategic stakeholder engagement, transparent communication, and innovative practices that balance profit with purpose, businesses are positioning themselves as leaders in sustainability, setting new standards for success that prioritize ecological stewardship and social welfare alongside financial profitability. This chapter emphasizes the transformation of businesses into entities that not only respond to but also anticipate and drive societal change toward a sustainable future.

# Chapter 12:
# Ethics in Marketing and Advertising

*Ethics in Marketing and Advertising* emphasizes the critical role of ethical marketing and advertising in promoting sustainability. It highlights the importance of transparency and responsibility in shaping consumer perceptions and aligning with their environmental values. The chapter warns against greenwashing, advocating for authenticity in how companies communicate their sustainability efforts. Purpose-driven marketing is presented as essential, not just for compliance but as a strategic approach to foster deep connections with consumers, encouraging them to make sustainable choices. Overall, the chapter calls for marketing strategies that genuinely reflect a company's commitment to positive environmental and social impact.

# Chapter 13:
# Measuring Impact:
# Metrics and Standards for Ethical Business

*Measuring Impact: Metrics and Standards for Ethical Business* explores the crucial role of metrics and standards in assessing the sustainability efforts of businesses. It emphasizes the importance of Environmental, Social, and Governance (ESG) performance as integral to a company's operations, highlighting the development of Sustainability Key Performance Indicators (KPIs) that align with global standards. The chapter advocates for robust, transparent metrics that refute greenwashing and enhance accountability, ensuring that every sustainability claim is measurable and verifiable. It stresses the future of sustainability reporting as a significant tool for ethical decision-making, pushing businesses beyond mere compliance to active custodians of impactful practices that foster a sustainable and equitable future.

# Chapter 14:
# Government Policies and Corporate Accountability

*Government Policies and Corporate Accountability* explores the dynamic between government policies and corporate accountability, emphasizing how regulatory frameworks not only enforce compliance but also stimulate corporate responsibility towards sustainability. It discusses how well-designed regulations can act as catalysts for businesses to adopt more sustainable practices, and the importance of aligning these policies with corporate strategies through public-private partnerships. The chapter urges corporations to advocate for policies that prioritize planetary and social welfare, positioning them as proactive participants in shaping a sustainable future. It underscores the potential of government policies to transform the landscape of sustainable business practices by encouraging corporations to integrate ecological and social responsibility into their core operations.

# Chapter 15:
# The Role of NGOs and Universities in Shaping Business Sustainability

*The Role of NGOs and Universities in Shaping Business Sustainability* delves into the pivotal roles that NGOs and universities play in shaping corporate sustainability. NGOs influence ethical policy development and implementation with their grassroots insights, while universities contribute through research and the cultivation of future leaders. Together, they drive businesses beyond profit-focused models towards sustainable practices, serving as educators, collaborators, and innovators. This chapter highlights how these institutions not only critique and improve corporate behaviors but also partner directly with businesses to embed sustainable practices deep within their operations, ultimately guiding companies through the evolving terrain of corporate responsibility.

# Chapter 16:
# Sustainable Finance: Investing in a Better Future

*Sustainable Finance: Investing in a Better Future* explores the transformative impact of sustainable finance, highlighting its role in driving ethical business and investment practices. As financial mechanisms evolve, green bonds and impact investing emerge as pivotal tools, encouraging investors to align their capital with environmental stewardship and social equity. This new financial paradigm challenges traditional investment models by integrating Environmental, Social, and Governance (ESG) criteria, thereby redefining profitability to include responsible, sustainable growth. The chapter argues for a holistic approach where financial success is intertwined with the global community's well-being, emphasizing that investing in sustainability is essential for long-term prosperity and environmental health.

# Chapter 17:
# Education for Sustainability:
# Raising Awareness and Building Skills

*Education for Sustainability: Raising Awareness and Building Skills* emphasizes the critical role of education in fostering sustainability within businesses and broader society. It advocates for integrating sustainability deeply into educational curricula and corporate training programs, not just as an added component but as a fundamental ethos. The chapter outlines strategies for developing sustainability-focused curricula in schools and continuous professional development programs in businesses, aiming to equip individuals with the necessary skills to innovate and lead in sustainable practices. By cultivating a mindset that prioritizes long-term ecological integrity and social equity, education for sustainability prepares individuals to tackle future challenges and drive meaningful change.

# Chapter 18:
# The Green Workplace:
# Creating Sustainable and Healthy Environments

*The Green Workplace:*

*Creating Sustainable and Healthy Environments* explores the transformative approach to creating sustainable and healthy work environments, emphasizing the integration of green design and practices within corporate spaces. It discusses how reimagining workplace settings to be more energy-efficient and employee-friendly not only enhances well-being and productivity but also significantly lowers carbon footprints. The chapter highlights the importance of biophilic design, the use of sustainable materials, and the promotion of green policies among employees to foster a culture of environmental responsibility. By doing so, businesses not only contribute to global sustainability efforts but also improve operational efficiency and employee satisfaction, setting new standards for what a healthy, sustainable workplace should embody.

# Chapter 19:
# Leadership for a Sustainable Future

*Leadership for a Sustainable Future* delves into the critical role of leadership in steering organizations towards sustainable futures, emphasizing the need for ethical reformation at all levels of governance. It outlines the qualities essential in leaders to foster a culture of sustainability, such as ethical reasoning, empathy, and a commitment to long-term ecological and social goals. The chapter highlights how leaders can inspire collective action and drive transformative change by embedding sustainability into corporate strategy, engaging stakeholders, and promoting transparency. Through education, leadership commitment, and innovative practices, it presents a roadmap for organizations to align their operations with sustainable and ethical standards, thus ensuring their contributions to a more sustainable world.

# Chapter 20:
# Entrepreneurial Solutions to Environmental Challenges

*Entrepreneurial Solutions to Environmental Challenges* delves into how entrepreneurs are pivotal in addressing environmental challenges by integrating sustainable innovations into business models. It highlights the concept of social entrepreneurship, where the main goal is to achieve significant social and environmental change. These entrepreneurs not only contribute to ecological preservation but also influence larger corporations and policy-making, setting new standards for sustainability. The chapter discusses challenges like scaling and funding, noting emerging solutions like impact investing. It underscores the importance of education in preparing future entrepreneurs to sustainably innovate, advocating for a shift towards business strategies that incorporate sustainability at their core.

# Chapter 21:
# The Rise of Conscious Capitalism

*The Rise of Conscious Capitalism* explores the shift in business practices towards recognizing the interconnectedness of economic success and societal accountability. This paradigm, influenced by the book "A Planet in Balance," highlights the need for ethical and socially conscious business practices. It describes conscious capitalism as a framework where businesses act as global community members prioritizing environmental preservation, social justice, and human rights. The chapter presents a vision for businesses as custodians of societal and ecological well-being, advocating for a shift beyond traditional metrics of success towards practices that foster community resilience and sustainable supply chains. This approach aims to guide businesses towards being leaders in a new era of ethical commerce, emphasizing the synergy between economic vitality and global stewardship.

# Chapter 22:
# Circular Economy:
# Redefining Growth and Consumption

In *Circular Economy: Redefining Growth and Consumption* the circular economy is presented as a sustainable alternative to traditional growth and consumption models, focusing on resource efficiency and waste reduction. This model encourages businesses to rethink product life cycles, aiming to keep materials in use for as long as possible and regenerate natural resources. The chapter outlines practical applications for businesses, such as adopting Product-as-a-Service models and enhancing material innovation. By integrating these circular principles, companies can not only mitigate their environmental impacts but also unlock new opportunities for growth and innovation, paving the way for a sustainable economic future.

# Chapter 23:
# Global Partnerships for Sustainable Development

*Global Partnerships for Sustainable Development* delves into the significant role that global partnerships play in achieving sustainable development. It underscores the effectiveness of collaborative efforts across borders and sectors, including governments, NGOs, businesses, and communities, in tackling global sustainability challenges. These partnerships leverage diverse resources and expertise, aligning corporate strategies with the Sustainable Development Goals (SDGs) to create impactful, long-term solutions for social, economic, and environmental issues. The chapter highlights the necessity of cross-sector alliances like public-private partnerships (PPPs), which combine governmental oversight with corporate efficiency to foster innovative and sustainable infrastructure projects. It stresses the importance of inclusive cooperation that integrates local knowledge into global initiatives, promoting a unified approach to sustainability that respects cultural differences and harnesses technological advancements for greater accountability and efficiency in sustainability projects.

# Chapter 24:
# Climate Change Mitigation:
# The Business Response

*Climate Change Mitigation: The Business Response* focuses on how businesses are responding to climate change by shifting their operational models and core values towards sustainability. Recognized initially as major contributors to environmental degradation, corporations are now transforming into proactive stewards of the environment. They're adopting innovative strategies to minimize their carbon footprints, integrating renewable energy sources, and reevaluating supply chains to ensure sustainability. The chapter underscores the importance of corporate governance reform, emphasizing not just regulatory compliance or enhancing brand image, but essential survival, resilience, and alignment with global sustainability movements. Through these transformative actions, businesses are not only participating in environmental preservation but are also driving significant advancements in the global response to climate change.0

# Chapter 25:
# The Future Horizon:
# Ethical Business in a Transforming World

*The Future Horizon: Ethical Business in a Transforming World* serves as the concluding chapter of the book, presenting a forward-looking vision where ethical business practices are central to navigating the complexities of a changing global landscape. This chapter underscores the pivotal role of businesses in addressing climate crises, societal inequities, and the challenges posed by disruptive technologies. It portrays businesses as stewards of sustainability and ethics, urging them to steer their strategies towards social responsibility and environmental preservation. Highlighting the growing expectations for businesses to act as proactive agents of change, it calls for the integration of sustainability deep into corporate operations and values. As the final chapter, it challenges companies to redefine corporate success, advocating for a future where business practices are in harmony with planetary well-being and human dignity, setting a new standard for ethical business in a transforming world.

# Embracing Our Responsibility and Opportunity

As this book concludes with *Embracing Our Responsibility and Opportunity*, we reflect on the transformative journey towards integrating sustainability and ethics in business. It's not merely about compliance; it's about embracing a profound shift that aligns business operations with the imperatives of planetary stewardship and social equity. This journey reshapes how businesses view success, urging them to prioritize long-term impacts over immediate profits and embrace diversity and inclusion to enrich their sustainability efforts. As we look forward, every decision and strategy must contribute to a sustainable future, demanding not just commitment but actionable change from businesses and individuals alike. This final chapter is a call to action, urging all to lead with responsibility and innovate for a thriving, equitable world.

James Fountain

# Glossary of Terms in *Our Changing World*

Throughout our conversation on *Our Changing World* and the multifaceted dimensions of sustainability, we've encountered a lexicon that is as varied as it is vital – a language shaping our understanding of not just where we stand, but also where we must go. This glossary serves as a compass to navigate the terrain of sustainability terms that appear throughout the book, ensuring clarity and fostering a deeper appreciation for the urgency and beauty in this journey toward a sustainable future.

### Acculturation
Acculturation is a complex and multifaceted process where individuals or groups from one culture engage in continuous and direct interaction with another culture. This process leads to the adoption of new values, behaviors, and practices from the encountered culture.

### Adaptation
The process through which individuals, communities, and ecosystems adjust to changes in their environment to mitigate harms or exploit beneficial opportunities. It is a cornerstone for resilience in the face of climate change, adjusting to its inevitable impacts.

### Albedo
A measure of how much light that hits a surface is reflected without being absorbed. Surfaces with high albedo, such as ice and snow, can reflect more sunlight and affect climate by keeping temperatures cooler,

whereas surfaces with low albedo, like forests or oceans, absorb more sunlight and can contribute to warming.

## Animism
A belief system that posits that all objects, places, and creatures possess a distinct spiritual essence. In sustainability, this perspective can influence environmental policies and conservation efforts, emphasizing the intrinsic value of nature and the need to respect and preserve the spiritual integrity of the natural world.

## Anthropocene
A proposed epoch marking the significant global impact of human activity on the Earth's geology and ecosystems, including climate change and biodiversity loss

## Agroforestry
The practice of integrating trees and shrubs into crop and livestock farming systems to increase biodiversity, improve soil health, and enhance ecosystem services

## Biodiversity
The variety within and among all species of plants, animals, and micro-organisms and the ecosystems of which they are part. This includes diversity within species, between species, and of ecosystems, forming the web of life of which humans are an integral part and upon which they fully depend.

## Carbon Footprint
A measure of the total amount of greenhouse gases produced to directly and indirectly support human activities, typically expressed in equivalent tons of carbon dioxide ($CO_2$). Reducing our carbon footprint is key to combating climate change.

## Circular Economy

An economic system aimed at minimizing waste and making the most of resources. This regenerative system aims to close the gap between production and natural ecosystem cycles—a stark contrast to the traditional linear economy, which has a 'take, make, dispose' model of production.

## Climate Change

A long-term change in the average weather patterns that have come to define Earth's local, regional, and global climates. These changes have a broad range of observed effects that are synonymous with the term global warming.

## Conservation

The protection, preservation, management, or restoration of natural environments and the ecological communities that inhabit them. Conservation is a means to ensure that nature will be around for future generations to enjoy and also recognizes the integral role nature plays in providing ecosystem services.

## Corporate Social Responsibility (CSR)

A framework for businesses to voluntarily integrate social and environmental considerations into their operations and interactions with stakeholders. It's a commitment to manage the economic, social, and environmental impacts of a company's operations responsibly and in line with public expectations.

## Cultural Capital

The collection of symbolic elements such as skills, tastes, posture, clothing, mannerisms, material belongings, and credentials that one acquires through being part of a particular social class.

## Cultural Homogenization

The process by which local cultures are assimilated and eroded by dominant cultures, often as a result of globalization. This can lead to a loss of cultural diversity and the unique knowledge and practices that contribute to sustainability and resilience in various environmental contexts.

## Cultural Sensitivity

Awareness and respect for cultural differences and the willingness to understand, communicate with, and effectively interact with people across cultures.

## Cultural Sustainability

Maintaining and evolving cultural beliefs, practices, and heritage as part of a community's overall sustainability goals, with respect for diversity and tradition.

## Cultural Tipping Points

Moments when a cultural norm or practice reaches a threshold and spreads rapidly, which can have significant implications for social sustainability.

## Eco-Friendly Apparel

Clothing made from organic or recycled materials, using processes that minimize the environmental footprint during production, distribution, and disposal.

## Eco-efficiency

Eco-efficiency is achieved by delivering competitively priced goods and services that satisfy human needs and bring quality of life, while progressively reducing ecological impacts and resource intensity throughout the life-cycle, to a level at least in line with the Earth's estimated carrying capacity.

## Ecological Footprint

A measure of how much area of biologically productive land and water an individual, population, or activity requires to produce all the resources it consumes and to absorb the waste it generates.

## Ecosystem Services

The many and varied benefits to humans that are provided by the natural environment and from healthy ecosystems. These include, but are not limited to, provisioning, regulating, cultural, and supporting services that directly or indirectly benefit human well-being.

## Environmental, Social, and Governance (ESG)

Environmental, Social, and Governance (ESG) criteria are a set of standards for a company's operations that investors use to screen potential investments. Environmental criteria consider how a company safeguards the environment; social criteria examine how it manages relationships with employees, suppliers, customers, and communities; governance deals with a company's leadership, executive pay, audits, internal controls, and shareholder rights. ESG examines a company's non-financial materiality that may influence investor engagement.

## Environmental Stewardship

The responsible management and care of the environment and natural resources with an emphasis on preserving and enhancing biodiversity and ecological integrity.

## Epistemological Diversity

Recognition of the existence of multiple ways of knowing and understanding the world, which can be influenced by culture, language, and personal experience.

## Ethical Consumerism

The practice of purchasing products and services that are produced ethically, considering the labor conditions, environmental impact, and animal welfare. Ethical consumerism encourages sustainable production practices and corporate social responsibility, influencing market trends toward more sustainable options.

## Feedback Loop
A system where the output of a process is used as an input, leading to further output that may amplify (positive feedback) or diminish (negative feedback) the process.

## Geopolitical Tensions
Political tensions influenced by geographic factors, often related to resource conflicts or environmental impacts.

## Green Economics
An economic framework that takes into account ecological and social costs, promotes sustainability, and values the well-being of both the environment and society.

## Grassroots Movements
Local, community-driven movements that grow to influence larger populations and policies, often associated with sustainability.

## Holistic Thinking
An approach that considers the interconnectedness and complexity of systems, crucial for addressing sustainability challenges.

## Indigenous Knowledge
Local knowledge unique to a culture, often contrasting with global knowledge systems and important for sustainability.

## Indigenous Land Rights
The recognition of Indigenous peoples' rights to their traditional lands and resources, essential for cultural preservation and sustainable practices.

## Interconnectedness
The recognition of the dependence of all life forms and ecosystems on each other, leading to an understanding that actions taken in one area can have global implications.

**Intergenerational Equity**
The concept of fairness or justice in relationships between the present and future generations, particularly in terms of resource allocation.

**Isolationism**
A policy or doctrine of isolating one's country from the affairs of other nations by declining to enter into alliances, foreign economic commitments, international agreements, etc., seeking to devote the entire efforts of one's country to its own advancement and remain at peace by avoiding foreign entanglements and responsibilities. In sustainability, isolationism can hinder global collaborative efforts needed to address worldwide environmental and social challenges.

**Moral Imperatives**
Principles or ethical considerations that compel individuals or societies to act in accordance with what is considered right and just.

**Multicultural Curriculum**
Educational syllabi that incorporate diverse cultural perspectives and content, thereby promoting inclusivity and understanding among students of different backgrounds.

**Regenerative Design**
Regenerative Design is a process-oriented approach to design. The goal is to develop systems that are capable of regenerating or restoring their own sources of energy and materials, thus creating sustainable patterns of consumption and production.

**Renewable Energy**
Energy from sources that are not depleted when used, such as wind or solar power, which are essential for sustainable development.

**Resilience**
The capacity of a system, community, or society potentially exposed to hazards to adapt, by resisting or changing in order to reach and maintain

an acceptable level of functioning and structure. This is determined by the degree to which the social system is capable of organizing itself to increase its capacity for learning from past disasters for better future protection and to improve risk reduction measures.

## Social Cohesion
The willingness of members of a society to cooperate with each other in order to survive and prosper, which can be essential for sustainable development.

## Social Resilience
The ability of a community to withstand external shocks and stresses as a result of social capital and community resources.

## Social Equity
The fair and just treatment of all individuals within society, ensuring equal access to opportunities and resources, and the protection from discrimination.

## Societal Equilibrium
A state of balance in a society or ecosystem, where the social or natural structure is maintained over time, often through sustainable practices.

## Socioeconomic Resilience
The ability of a social and economic system to recover from shocks and stresses, such as economic crises or natural disasters.

## Sustainability
The ability to meet the needs of the present without compromising the ability of future generations to meet their own needs. Sustainability is often broken into three pillars: environmental, economic, and social, also known informally as planet, profit, and people.

**Sustainable Agriculture**
Farming that meets the needs of the present without compromising the ability of future generations to meet their own needs, typically involving environmentally friendly practices.

**Sustainable Development**
Development that meets the needs of the present without compromising the ability of future generations to meet their own needs, encompassing a balance between environmental, economic, and social goals.

**Systems Thinking**
A holistic approach to analysis that focuses on the way a system's parts interrelate and how systems work over time within the context of larger systems.

**Tipping Point**
A critical threshold at which a small change or influence can lead to a significant and often irreversible effect on a system. In sustainability, tipping points are crucial in the context of climate change and biodiversity, where they represent points beyond which systems may not recover, leading to drastic changes in the environment.

**Traditional Ecological Knowledge (TEK)**
A cumulative body of knowledge, practice, and belief, evolving by adaptive processes and handed down through generations by cultural transmission, about the relationship of living beings (including humans) with one another and with their environment.

**Tragedy of the Commons**
A situation in a shared-resource system where individual users acting independently according to their own self-interest behave contrary to the common good of all users by depleting or spoiling that resource through their collective action.

**Zero-Carbon**

Referring to an operation or activity that releases no carbon dioxide into the atmosphere. This ambitious goal can be approached by reducing emissions and implementing carbon offset schemes that compensate for any emissions that are produced. Zero-carbon is a guiding star in the journey toward climate neutrality.